THYLACINE

CONFIRMING TASMANIAN TIGERS STILL LIVE

ALAN HEATH

Figure 1 An example of very dense rainforest on a roadside in North Queensland, with back of a vehicle to show scale. A very good hiding place for a Thylacine-size animal. Author's collection.

Published by Vivid Publishing
P.O. Box 948, Fremantle
Western Australia 6959
www.vividpublishing.com.au

National Library of Australia Cataloguing-in-Publication entry
Creator: Heath, Alan R. (Alan Richard), author.
Title: Thylacine : confirming Tasmanian tigers still live / Alan Heath.
ISBN: 9781925209402 (paperback)
Subjects: Thylacine--Tasmania.
Thylacine--Australia.
Extinct animals--Tasmania.
Extinct animals--Australia.
Dewey Number: 599.27

Cover photograph: Tasmanian Archive and Heritage Office:ph30/1/5847A
 Revised second edition.

DEDICATION

To my parents, Bill and Amy, who nurtured in me a love of the natural world.

ABOUT THE AUTHOR

Born in Fulham, London, Alan Heath lived the first seven years of his life in Fulham, and the next four years in Saint Mary Cray, Kent. He came to Sydney by ship as an almost eleven year old with his parents and four siblings in early 1955 as part of the immigration scheme that was then in place. The first five and a half years or so in Australia he spent at Bunnerong Migrant Hostel, at Matraville which is in the south eastern suburbs of Sydney, until the family's new home was completed at Collaroy Plateau on the northern beaches of Sydney.

He 'followed in his father's footsteps,' serving an apprenticeship, and became a Licenced Electrician. After gaining his Electricians licence he worked for two years as an Electrician on the Snowy Mountains Hydro-Electric Scheme in southern New South Wales, He then returned to live in Sydney and continued to work in the electrical industry. In 1974 he completed the Electrical Engineering Certificate course, passing it with honours. After retiring as an Engineering Officer in the electrical industry he moved to Queensland, where he has lived since 2004.

He has always had an interest in engineering, history, natural history, and photography, and as an adult he has developed an interest in geology, palaeontology and anthropology.

CONTENTS

PREFACE

I first learnt of Tasmanian Tigers on radio news bulletins soon after arriving in Australia. Because I did not want this unique animal to become extinct, I was always heartened to hear these bulletins which stated that this very rare animal had been seen in Tasmania. I am still heartened when I hear bulletins of them being sighted. When I heard about them for the first time, I assumed that they were the same as the tigers that are found in other parts of the world, and I wondered why, in England, I had not heard about tigers also living in Tasmania. I never dreamt that one day I would be on a quest to find out what people knew about the existence of them, both in Tasmania and on the mainland of Australia.

In this book, which gives only the basic facts, I have recorded some of what people, who I believe are truthful, have told me about what they know about living Thylacines. To minimise any disturbance of the animal I have omitted some of what I have been told. I have also not disclosed the workplaces of at least two people who I have mentioned in this book. Initially, I only recorded what they had told me for my own information and education.

When I started out in 1993 trying to find people who knew about living Thylacines I had no idea that I would meet so many who were happy to tell me the details of what they, one of their truthful relatives, or one of their friends had told them about seeing them.

I have written this book for the benefit of Zoologists and other researchers, and for anyone who is interested in this unique animal. I hope that it helps them in their Thylacine research and studies.

The Thylacine has the zoological name of Thylacinus Cynocephalus, which means pouched dog with a wolf's head.

ACKNOWLEDGEMENTS

I sincerely wish to thank all of the people who have freely told me about what they know about living Tasmanian Tigers. Without their information I could not have written this book. I especially want to thank the Cooktown resident who told me in 1993 about his sighting in the Iron Range on Cape York Peninsula. You made me realise that there is hope for the survival of the Tasmanian Tiger on mainland Australia.

I also want to sincerely thank the Zoologist who told me in 1993 that he and his colleagues had made plaster castes of Thylacine footprints at various locations on the mainland of Australia. He initiated in me a desire to find out from people living in rural areas on the mainland, and in Tasmania, what they knew about living Thylacines.

INTRODUCTION

I loved being taken by my parents to the London, Chessington, and Whipsnade Zoos with my brothers and sisters when we lived in England. I also loved being taken by dad with my brothers to The Natural History Museum in London, where there was much to see in the way of fantastic animals etc., both living and extinct. I must admit though that my favourite museum was the nearby Science Museum, because of the numerous electrically-driven models of steam-driven water pumps etc. as well as all of the other original trains etc. Sometimes dad would let us children have a vote on what museum we wanted to visit for the day.

When I arrived in Australia, I found that there were many more animals here than the Kangaroos, Wallabies, poisonous snakes and spiders that all British people seemed to know about, and the latter two feared. To me as a child it seemed that the vast bushland areas of Australia were what we would now call an open-range zoo, which was completely different to England.

The reader may think that this is a strange, or even a silly question, but, were Captain James Cook and Joseph Banks the first people to record the existence of Thylacines on the Australian mainland at what is now Cooktown in the far north of Queensland in 1770? I believe that they were, because, as you will read in the following pages, people have told me of them living in the Cooktown area, and in all parts of Cape York Peninsula. Very few people would have read their journals, but I was lucky enough to purchase both of them from second-hand bookshops in 1978 and 1994 respectively, and I finished reading each of them during the year that I bought them. I can recommend them to anyone who would enjoy reading historical adventure books. They are titled: 'The Journals of Captain James Cook the Voyage of the Endeavour, 1768-1771.' Edited by J.C. Beaglehole, and, by the same editor, 'The Endeavour Journal of Joseph Banks 1768-1771.'

J.C. Beaglehole was a historian at Victoria University of Wellington in New Zealand, and in these journals he copied exactly what was written in the originals, spelling errors included. He also made numerous explanatory notes so that the modern reader could better understand details recorded in the originals.

Cook was instructed to record full details of the peoples and the wildlife etc. that he found at all of the places that he visited during the voyage, because this was to be a voyage of scientific research, and he did this zealously.

On the 4-8-1770 Cook recorded in his journal the animals that were to be found at what is now, Cooktown. Two of the animals that he recorded were a tame dog, which would often come around the camp to be fed by the people of the 'Endeavour'. The other animal that he recorded was 'Wolves.' Beaglehole makes the notes that the dog would have been a Dingo, and the 'Wolves' may have been a Thylacine, because of it's superficial similarity to a wolf. Thylacines were known as 'Marsupial Wolves' in the early 1900s because of this similarity.

On 29-7-1770 Joseph Banks recorded in his journal that an American midshipman had told him that he had shot at a wolf that 'was perfectly the same as those that he had seen in America', but he had not killed it. There was only one American midshipman on the 'Endeavour,' and he was James Magra, who later changed his surname to Matra. The suburb of Matraville in Sydney is named after him.

Note: In 1770 'America' was what is now Canada and the United States of America.

I can understand why Magra wanted to kill the 'Wolf,' because there was live food such as pigs etc. placed on shore while the ship was being repaired. Magra would have thought that he was doing his duty in trying to kill it, as well as defending himself and his shipmates from what he believed was a large dangerous predator.

I believe that Magra would assume that the animal that he shot at was a wolf because of its very large size compared to the Dingo that he had no doubt seen in their camp. He possibly only saw the head and shoulders of the 'wolf' protruding from the long grass etc. so he would not have seen the stripes running across it's back. Shadows from the vegetation may have also obscured them.

I had a real problem understanding how Magra, who obviously was familiar with the wolves living in North America, could tell Banks that the wolf that he had shot at that day looked 'perfectly the same' as the wolves that he had seen in America, because all photos and movies of wolves that I had ever seen showed that they have long hair, while a Thylacine has short hair. This problem arose because of my lack of knowledge about wolves, but it was solved after I had a talk to a man from New Zealand in approximately 2008. He told me that the year before he went to Canada for a holiday, and while there he, and his wife, had seen short-haired wolves.

I did not disbelieve this man, but I felt that I should confirm this by contacting the head office of the Canadian national parks service. I did this and I received an e-mail in reply stating that wolves shed their long winter coat in spring.

Note: I did not tell the Canadian national parks officer that I wanted to know if wolves have short hair like that of Thylacines.

I believe that the question that has to be answered is, what Australian animal was seen and recorded as being a wolf, when some of the people on the 'Endeavour' were feeding the tame Dingo, and most, if not all, would have seen it around their camp? I believe that the only logical answer to this question is, a Thylacine.

Tasmanian Archive and Heritage Office: ph30/1/5847A

Figure 2 Statue of Captain James Cook in the park on the bank of the Endeavour River at Cooktown. Author's collection.

Figure 3 Statue of Captain James Cook at Cooktown, with the Endeavour River entrance in the background. Author's collection.

Figure 4 Looking North West across the entrance of the Endeavour River to Mt. Saunders from below Grassy Hill, Cooktown. Author's collection.

Figure 5 Cooktown, taken looking South West from the top of Grassy Hill, which Captain Cook and some of his crew climbed to try to find an escape route through the reef in 1770. Author's collection.

Figure 6 Endeavour River at low tide, Cooktown, from Grassy Hill. Author's collection

Figure 7 The 431m, (1415foot), high Mount Cook on the south side of Cooktown, from top of Grassy Hill. Author's collection.

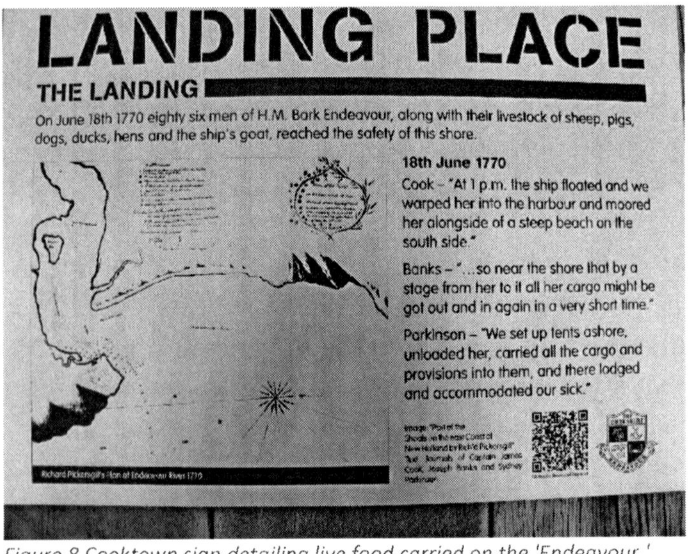

Figure 8 Cooktown sign detailing live food carried on the 'Endeavour,' Cook's original chart, and three 1770 journal entries. Author's collection.

The seeds of my quest for Thylacine information were sown in late 1993, when a Zoologist working in an educational institution told me that he was a fully-qualified Zoologist who had expert knowledge of Australian animals. He also told me that he and his colleagues had taken plaster castes of Thylacine footprints at various places on the mainland of Australia. He told me that he and his colleagues know that these footprints were not man-made hoax footprints, so I assume that they were found in isolated muddy areas where the only human footprints in the vicinity, were their own. He also told me that they have received written reports, which they keep in their files, from all parts of mainland Australia from Thylacine witnesses.

I spoke to this Zoologist by phone very soon after returning from a holiday in north Queensland, where during my first visit to Cooktown I met a local man who told me, and several other holiday makers, who were from overseas, that he had seen the same Tasmanian Tiger several times one night as he camped alone with his dog in a very isolated and uninhabited area that was much further north than Cooktown.

The Zoologist asked me if I would be willing to send him a written report containing all of the details that I had just told him, because he and his colleagues search for Thylacine evidence in the areas mentioned in all written reports that they receive. I replied that I would be happy to do this because as far as I was concerned this is very important information for Zoologists. At the time I thought that Thylacines had been extinct on the mainland for a thousand years or more. He also told me that they had no reports of Thylacines being sighted in that Cape York area. I wrote the report and sent it to him a day or two later. Details about this sighting are given later in this book. The Zoologist also told me that over the years the average number of written reports his institution has received about Thylacine sightings on the mainland of Australia has been twelve per year.

After the Zoologist told me about the plaster castes that he had taken, I told him that he should inform all of the Zoologists world-wide that the Thylacine is not extinct. To my surprised he replied that this evidence would not be accepted by them. I then asked him why this evidence was not good enough for them, he replied, because they would want a live animal as evidence. I then realised that in the scientific world, only indisputable evidence is accepted as proof of their existence.

Because of what this Zoologist had told me about the plaster castes he and his colleagues had taken, I decided, whenever I could, to try to confirm that Thylacines exist on the Australian mainland, and also in Tasmania. I never thought that I would be told so consistently by complete strangers about what they knew about living Thylacines. In 1993 I had no intention of ever writing a book about what I may be subsequently told about living Thylacines.

Also, on the ABC radio program 'Bush Telegraph' website of past programs, there is an audio recording of a program that was broadcast on 6 September 2011. In this program Marie Attard, and Doctor Bob Paddle, who wrote the book 'The Last Tasmanian Tiger The History and Extinction of the Thylacine,' are being interviewed by Michael Cathcart about their Thylacine research. Doctor Paddle states in the audio recording that the first lists of animals that were drawn up by Naturalists and scientists from their states of New South Wales, Victoria, and South Australia in the 1840s, all recorded Thylacines as living in their state. Doctor Paddle also states that there were sightings, skins and bones of Thylacines in each of these states at this time. Note: In the 1840s Queensland did not exist because it was part of New South Wales.

Recently, after telling an interested academic at a University about some of the things that people have told me regarding living Thylacines, he told me that I should write a book detailing what they had told me. I replied that I did not want to do this because I do not want to be responsible for any accidents, or worse, that may happen to inexperienced people who were walking around in the remote and rugged areas of Queensland looking for Thylacines because of what they had read in my book.

After thinking about this, and because of the fear that a woman told me that she had of being thought of as being mad if she told anyone about her Thylacine sighting, I decided to write this book so that other people who may have this same fear of disclosing their Thylacine sighting are comforted to know that other people have also seen them. Hopefully these people will then want to disclose details of their sighting to the academic world, and possibly to the public. I do not believe that people should have to live in fear of being called mad or ridiculed, and probably ostracised from their community if they disclose that they have seen one.

To alert anyone who decides that they will look for Thylacines, I have decided to include a warning in this book. It lists some of the dangers when walking or driving in the bush of northern Australia, Some of these warnings are also applicable to other parts of mainland Australia, and to Tasmania.

I have never planned to search for Thylacines because they are a rare, shy, and mostly a nocturnal animal, and my chances of being lucky enough to see one would be very slim.

I have included some of the photographs that I have taken of Queensland forests to show the reader how thick, rugged, and vast they are, and how easy it would be for Thylacines to live in them without being seen. I have also included other photographs that I have taken of Cooktown, where Captain Cook and Joseph Banks recorded in their 1770 journals that 'wolves' were living there.

Alan Heath, 2014.

WARNING

I do not want the reader to think that I am in any way recommending, or, in any way encouraging people to search for Thylacines because of what I have written in this book. For the inexperienced, Cape York Peninsula, the northern, and any other isolated part of Australia are extremely dangerous places. You must be aware of the dangers of walking or driving in the bushland, or in the rainforests in these areas, some of which are:

The temperatures in northern Australia in summer are consistently very high, as is the humidity in the tropics. Without water there are areas where a person may die within a day.

Cyclones, which are also known as Hurricanes or Typhoons in other parts of the world, cross the coast of Queensland in the summer months i.e. usually from November to April, but sometimes before and after these months.

Several species of very dangerous stinging jellyfish, such as the Box Jellyfish, are to be found in the saltwater rivers and in the ocean in northern Australia.

There are vast areas with very few inhabitants, so if a person is lost or injured their chance of survival is very low. Families living in these areas have High Frequency radios which transmit and receive messages over thousands of kilometres. They also have satellite phones, because there is no mobile phone coverage where they live.

In the summer wet season the rivers can rise many metres suddenly, causing all roads to be cut for weeks or even months at a time. These rivers can also flow very swiftly, making them extremely dangerous to swim across.

Very dangerous Saltwater Crocodiles can also live in freshwater rivers more than one hundred kilometres from saltwater. I have been told by a cattleman living in northern Queensland that they have been known to walk up a dry riverbed for twenty kilometres to the next waterhole. These crocodiles live in all rivers north of approximately Gladstone, as well as in the ocean. Note: The construction of the Burdekin Falls Dam, which is well inland from Ayr, trapped them upstream, and now they live in it.

Some of the most dangerous snakes in the world are to be found in Queensland. Taipans are very fast, and they strike very quickly.

There are some areas with numerous disused open, and deep vertical mine shafts which may be difficult to see because of vegetation.

There are many wild pigs in numerous areas. These can grow to a very large size. Boars with very dangerous tusks have been known to wait for hours at the bottom of a tree that has been climbed by a person who is trying to escape from them.

I have been told by an elderly neighbour who has always lived in Queensland, that a pack of Dingos have been known to follow a person through the bush for a very long way, waiting for the person to lay down to rest. He told me that if ever I was walking alone in the bush with a pack of them following me, then I should make sure that I stayed on my feet, because if I was to lay down they would think that I was sick, weak, or injured, and they would then attack me.

It would be very easy to get lost if a person decided to leave a road or track without a detailed map, compass, or a GPS. The forests in Queensland can extend for many hundreds of kilometres in all directions. Some rainforests are almost impossible to walk through. The road between Coen and 'The Tip' of Cape York Peninsula passes almost entirely through flat forest covered terrain, so, to see the area around you, you would need to climb a tree, or have a high temperature-rated small camera-equipped 'drone' helicopter and it's associated equipment.

Australia is well known for bushfires, which can be started by lightning. The leaves of Eucalypt trees contain oil which burns very readily. 'Crown' fires race across treetops very quickly in strong winds.

To give the reader an idea of the vast area of Cape York Peninsula, the following are the approximate road distances, in kilometres, between towns, road houses and the road turn offs that a traveller will encounter, in sequence, when driving north from the Atherton Tableland town of Mareeba to 'The Tip' of Cape York. Note: 1km equals approximately 0.62 miles:

Mareeba to Mount Malloy: 40.
Mount Malloy to Mount Carbine: 28.
Mount Carbine to Bob's Lookout: 25.
Bob's Lookout to Palmer River roadhouse: 58.
Palmer River roadhouse to Lakeland: 31
Lakeland to Laura: 64.
Laura to Hann River roadhouse: 76.
Hann River roadhouse to Musgrave: 127.
Musgrave to Port Stewart turnoff: 82.
Port Stewart turnoff to Coen: 25.
Coen to Archer River roadhouse: 66.
Archer R. rh. to Portland Roads turnoff (south): 21.
Portland Roads t.o.(s.) to Portland Roads t.o.(n): 13.
Portland Roads turnoff (n) to Weipa turnoff: 14.
Weipa turnoff to Morton Telegraph Station: 72.
Morton Tele. Stn. to Bramwell Junc. roadhouse: 42.

Bram. Junction rh. to Capt. Billy Landing t.o.: 67.
Capt. Billy Landing turnoff to Jardine R. ferry: 106.
Jardine River ferry to Ussher Point turnoff: 23.
Ussher Point turnoff to Bamaga turnoff: 24.
Bamaga turnoff to Somerset turnoff: 23.
Somerset turnoff to car park at 'The Tip' beach: 15.

This is a total distance of approximately 1042km, or 647 miles. From just south of Mount Malloy the driver, except in the Mount Malloy area, Lakeland area, and Laura, continually travels on a road that passes through various types of mostly dense forest!

All of the road to Laura is sealed, but north of Laura it is almost entirely corrugated gravel. The only sealed roads north of Laura are those in the towns.

The 'inland' road from Mareeba to Cooktown is sealed, but the 'coastal' one, which originates at Mossman, is mostly unsealed corrugated gravel north of the Daintree River ferry. It is only suitable for 4WD vehicles in dry weather only. North of this river the road mostly passes through thick rainforest to Cooktown.

Figure 9 Road condition and crocodile warning signs at Lakeland on the road to Cape York. Author's collection

Figure 10 A crocodile warning sign, showing in black shading on a map of Queensland where they are found. Author's collection.

Figure 11 A 1.5m long extremely dangerous Taipan snake crossing a Queensland road. Author's collection

Figure 12 Tank outside the old Musgrave telegraph station on the road to Cape York. It shows road distances in kilometres. Author's collection.

METHODS I USED TO OBTAIN INFORMATION

I had to work out the best way of obtaining truthful information from complete strangers about what they may know about living Thylacines. To do this I generally used the following two methods:

The first method that I used was, when I was in Tasmania, was this: I decided that I must make people think that I was very sceptical about Thylacines still being alive. This, I hoped would make them to try to convince me that they were not extinct if they knew that they still existed. This method worked well.

In the rural areas where there was still a lot of forest, hills, or mountains, I would ask people of various ages above 25 to 30 years of age, and who I believed were sensible, 'I am from New South Wales, (or Queensland, as the case may be at the time), what is the truth about people living in Tasmania claiming that they have seen the extinct Tasmanian Tiger? I have heard these reports in the media since I first came to Australia in 1955, are they publicity hunters who are just trying to get their faces on TV, or their names in print in the newspapers?'

Approximately half of the fifteen or so people who I have asked this have replied that they do not know anything about them still being alive, but the others would then tell me about what they, an honest relative, an honest colleague, or an honest neighbour etc., know about them existing. I did not like doing the following, because I did believe them, but as an extra test of their honesty several times I have then said to the person, 'I think that you may be having me on, because they are extinct.' All of them immediately showed intense dislike of me in their faces, because I was really accusing them of lying to me. I believe that if they had been lying to me, their faces would not have shown this intense dislike. I would then tell them that I do believe them, and I would then thank them for telling me about them being alive. The intense dislike in their faces would then disappear.

Since moving to Queensland, the second method that I used was as follows:, Again, if I was chatting to a person who appeared to be sensible, and they mentioned during our conversation that they had lived on Cape York Peninsula, I would tell them in a sceptical tone of voice that, 'I had heard rumours that the extinct Tasmania Tiger lives there, did you ever hear these rumours when you lived there?'

Of the approximately twelve people who told me that they had lived on 'The Cape', only one person told me that they did not know anything about them living there. All of the other people told me that they know that they live there. I would then tell them that 'I have modern reproductions of Captain Cook's and Joseph Banks' 'Endeavour' voyage journals edited by J.C. Beaglehole, and in them they record that there were 'wolves,' at what is now Cooktown, in 1770, so you have to answer the question, what Australian animal looks like a wolf, and is the size of a large dog?' I would then tell them, 'the only logical answer must be a Thylacine, because it is that size, and it looks like a wolf'.

Another method that I use in Queensland is this: If I meet a 'weather-beaten' man who looks like he may work in the cattle industry, or as a farmer etc. in isolated rural areas, I tell him that I have heard rumours of the extinct Tasmanian Tiger being sighted on Cape York Peninsular, and elsewhere in Queensland, and also that both Captain Cook and Joseph Banks recorded in their 1770 journals that there were wolves at what is now Cooktown. I would then wait for them to tell me anything that they may know about them. This method has also worked well in obtaining Thylacine sighting information.

If the reader has any doubt about the truth of the Thylacine-sighting information that I have recorded in this book, then I suggest that you make your own enquiries using the same methods that I have used. Then you will be able make up your own mind about the truth of what you have been told.

CHAPTER 1 THYLACINES IN QUEENSLAND

A HOLIDAY TO REMEMBER – IRON RANGE

In November 1993 I was holidaying in the tropical coast area of north Queensland, and I visited Cooktown for the first time.

For the last full day of my four or five-day stay in Cooktown, which is on the east coast of Cape York Peninsular, I decided to go on a full-day tour which included a tour of the Aboriginal art gallery site at Split Rock near the town of Laura, because I had never seen one of these galleries except on television.

This tour was conducted by a man who appeared to me to be a serious and methodical person who had lived in the remote parts of northern Queensland all of his life. He was approximately thirty five years old. Before the seven of us, who were to be taken in his four-wheel drive vehicle for the day set off, he told us that he was interested in both the European and the Aboriginal histories and cultures of Cape York Peninsula.

After viewing the art gallery we had lunch outside the heritage-listed original 'Laura' homestead, which is within Lakefield National Park, about 30km north of Laura. During this lunchbreak, a lady who was on holiday from Canada, asked the tour operator to tell her about the way that we live in Australia and about our lifestyle, because this was her first visit here. The rest of us all listened in silence to everything that he told us. She asked him about his job of driving tourists around, and then she asked him about his previous job. He replied that ‘last year I was way up the Cape from here, fossicking for gold for a year.’

He also told us that where he was fossicking was a very isolated and lonely area with very thick forest, and nobody lives anywhere near that area, so he had a bitch Alsatian dog for company. The lady then asked him about the way that he obtained the gold, and she also asked if he lived in a tent and how he cooked his food. He then truthfully replied to all of her questions, describing his method of panning, how he had a swag to sleep in, and how he did all of his cooking, bread included, in a camp oven. She asked him what a camp oven was so he described it to her, and how he cooked in it.

He then suddenly told us that one dark evening while cutting up the food to place in the camp oven, his bitch dog suddenly became hysterical and barked continuously. He had no idea what was upsetting her, because she had never behaved like this before. He told us that he thought that she was having a fit. He told us that he ‘could not shut her up’ as she continually looked into the forest in one direction, so he tied her up to a tree. He told us that he looked in the direction that she was, but could not see, or hear anything at all, so he took his torch and walked into that part of the forest. After walking almost fifty metres in a straight line, he saw in front of him in the distance ‘a big dog-like animal that had a large head and large shoulders, and looked different to the very numerous Dingos and wild dogs that he had seen.’

He told us then that he was thinking to himself, what is this strange looking big dog, because he thought that he knew all the native animals in the bush, but he didn’t know what this animal was? He told us that he was intent on getting as close as he could to it so that he could get as good a look at it as possible before it bolted away, so he very slowly walked towards it, keeping the torch shining on it all of the time. He told us that when he got nearly twenty metres from it, the animal, which he told us was as big as a large Alsatian dog, turned around and walked away from him. He told us that as it turned around he then saw black stripes running across it's back, and he thought 'what on earth is this animal?'

Note: I immediately thought that it was a Tasmanian Tiger, but I had only heard of them being in Tasmania, because they were believed to be extinct on the mainland of Australia for more than a thousand years. I did not admit to the man during the day that I knew anything at all about Thylacines, because I thought that this man must be the only man to have ever seen one on the mainland, and I was determined, if I could, to find out exactly where he saw it, so that when I got back to Sydney I could tell a fully-qualified Zoologist in a university, or Taronga Park Zoo etc. about what this man was telling us. As mentioned in the Introduction, at the request of the Zoologist that I contacted, I sent to him a written report about what this man told us.

The man told us that after turning around it walked away from him in a straight line with its head turned back frequently so that it could see what he was doing. The man told us that he followed it, keeping nearly 20m behind it, for nearly 100 m, then he sped-up his pace so that he could get a closer and better look at it, but it immediately also sped-up it's pace, so the man slowed back down to his original pace, and the 'big dog' also did the same.

He told us that 'it was happy for me to stay nearly 20m behind it.' He said he followed it for about another

500m in the same direction, keeping nearly 20m from it, until he decided to return to his camp to finish preparing his meal, because he thought that he would be following it all night otherwise. He told us that he was thinking, 'I wonder if it had been planning to attack my dog, and she knew the danger that she was in.' He told us that when he got back to his camp he found that his dog was quiet and 'back to normal,' so he finished placing his food into the camp oven.

Then approximately 20 minutes after returning to camp, and while his food was cooking, his dog suddenly re-commenced her hysterical continuous barking, but this time she was looking in another direction into the bush. Again he could not 'shut her up,' so again he tied her to the tree and then set off with his torch in the direction that she was looking, thinking that it must be back again. Again he walked for nearly 50 m, when he found the same 'big dog' facing him, so again he slowly walked towards it, so as to get the closest possible look at it. When he was nearly 20 m from it, it again turned around allowing him to see the stripes across its back, and started to walk away from him, frequently looking back at him. He told us that this second visit convinced him that it was trying to find a way past him so that it could attack his dog.

Again he followed if for nearly 100 m before he sped-up his paced, but again it also sped-up, so he again slowed down to his original pace and followed it for approximately 500 m, when he returned to his camp because his meal would be cooked when he got back to camp. On returning to camp his dog was again 'back to normal.'

He told us that after he had eaten his meal, his dog again started barking hysterically, and looking in a third direction into the bush, so he again did the same as he had done on the previous two visits. He found that it was the same 'big dog.' He told us that it did not return after it's third visit.

He told us that he was determined to find out what this 'big dog' with stripes across it's back was, so he decided that the next time that he met one of the old lone fossickers, who have been looking for gold on 'The Cape' for many years, he will ask him if he and the other old fossickers have ever seen them also. He told us that only people who live on 'The Cape' know that these old men are looking for gold there, because they are only seen when they come into a town for a day, ever six to eight week, to replenish their supplies, before returning to their search area, which is well and truly off of 'the beaten track.'

He later met one of these old men in a town, and he replied to his question that he, and the other old fossickers have seen them, but only very occasionally, but, he nor the others know what sort of a dog they are. He told us that he wanted to learn as much as he could about then, so he asked the old fossicker to tell him everything that he and the others know about them, to which he replied: 'We only see them on moonlit nights, because the bush is easier to see then. They wander into the bush near our campsites, and you can see them just standing there watching you, and in the moonlight the stripes across their back makes them look like a tiger. They stand there watching you for several minutes or so, and then they wander off into the darkness. They never attack you, but only want to look at you, and that is all that we know about them.'

The man then told us that he then made up his mind to report, to an animal expert at a National Parks Office, what he, and the old fossickers know about them, and to ask what their name was, and to find out everything else that he could about them, so the next time he was in a town with a NPS office he visited it and asked a man at the counter if there was an animal expert here. The officer replied that I'll bring him out to see you from the back office. When he came to the counter, the man told him that he wants to report seeing an animal that he has never heard of, or seen before, and to find out what it is called, because I thought that I knew all of the animals in Queensland. The officer said to him, 'tell me everything about what you saw, and I will tell you what it was.'

The man told the officer how his dog had become hysterical, otherwise he never would have known that the 'big dog' was there, and then he told him everything about his sighting, and about what the old fossicker had told him about them, as the officer listen in silence. When he had finished telling the officer everything, the officer replied, 'well mate, you have given me a perfect description of a Thylacine.' To which he replied, what sort of dog is that? The officer then explained to him what a Thylacine is, and told him that it was commonly known as a Tasmanian Tiger, where the last one died the 1930s, and they have been extinct on the mainland for thousands of years, so you have seen a Dingo. To which he replied, 'it definitely was not a dingo!

The officer refused to believe that he had seen a Thylacine, which annoyed the man, and then the officer accused him of making a false report, which made him more annoyed. He told us that he felt like punching the officer when he told him, 'you must have imagined seeing it.' The officer then told him to leave the office because he had seen a Dingo, so, in desperation he told the officer that he will drive him, and anyone else he nominates to the exact spot where I was that night, and return, because you may find evidence that Thylacines live there, and he will then know that he is telling him the truth. The officer declined his offer and told him to leave, which he did, extremely annoyed and insisting that he was telling him the truth about what he saw.

This was the end of what he told us about his sighting, and the other people in our group said, that that was an unusual story, and he then told us all to pack up because lunch was over now and we have to start on the afternoon part of the tour, but I was determined to find out, if possible, where he saw the Thylacine, because, when I got home I would find and tell a Zoologist what I had just been told. While the others started to pack up I walked over to the man and asked him quietly, 'where did you see that animal?, to which he replied, 'way up The Cape.' I thought that he does not want to tell me, so I tried another method of getting him to tell me. I then said to him in a quiet voice, 'but it's a vast area up there, it could have been anywhere, where did you see it?' He hesitated for approximately five seconds and then said somewhat reluctantly, 'It was at Iron Range.' I said quietly in reply, 'oh, right,' and then I started to also pack up my things.

I had never heard of the Iron Range, nor did I then know that there was gold north of the Palmer River, which is about 125km south west of Cooktown.

When I returned home to Sydney, I looked at my atlas to see if I could find the Iron Range on 'The Cape,' and there it was on the east coast, approximately due east of Weipa. This made me 99.99% sure that the man had been telling the truth, because he appeared to be an honest and serious man, and because I was, and still am, certain that he would never have told us about his sighting if the Canadian woman had not asked about our lifestyle. I then informed the Zoologist who I have mentioned in the Introduction, about this man's Thylacine sighting.

After hearing the man's story, I had, In the back of my mind a feeling that in 1978, when I read Captain James Cook's Journal, he recorded something about wolves at what is now Cooktown, so when I got home from this holiday I read in it where he recorded 'wolves' at Cooktown on 4 August 1770. This also convinced me that the man was telling the truth about what he had seen.

Tasmanian Archive and Heritage Office: ns463/1/8/4

Figure 13 The 400mm-deep Pascoe River crossing on the road to the Iron Range, and the hamlet of Portland Roads. Author's collection.

Figure 14 Iron Range, showing entrance to the Mt. Tozer lookout track, taken from the road to Portland Roads. Author's collection.

Figure 15 The Iron Range, showing the track to the Mt. Tozer lookout, and the very thick vegetation, in showery weather. Author's collection.

Figure 16 The Iron Range, showing the eastern rain forest-covered slopes of Mt, Tozer on a showery day. Author's collection

LOCKHART RIVER AREA

In 2009, as I travelled to Cairns, I met a friendly and an obviously intelligent man who was approximately thirty five years old. He was in the front garden of his beautifully restored old 'Queenslander' house, built on high posts, thus cooling it's interior. This is common in Queensland.

As we chatted about his house etc. he told me that he had worked for an extended period in a government department at three towns on Cape York Peninsula. I thought that 'government men' in very isolated areas would know a lot about what is happening in the areas where they are working, because word of things that happened there would be spread between the various government departments on the 'government grapevine.'

I thought that he would probably know that Thylacines live on 'The Cape,' so I told him that I had heard rumours from people who have lived there, that the 'extinct' Tasmanian Tiger lives on 'The Cape, and I then asked him if he had also heard about them living there. He instantly replied, 'They do live there!' I then replied in a surprised tone of voice, 'So these rumours are true then,' to which he replied, 'They are true.' I then asked him where on 'The Cape' they live, and again he instantly replied. 'They live at Lockhart River.' I then told him that I had heard that they live in the Iron Range, which is in that general area.

Note: I had not mentioned to him anything about Thylacines before I asked him if he had heard any rumours about them.

Figure 17 The road junction to the towns of Lockhart River, and Portland Roads, showing the thick rainforest of the area. Author's collection.

PORTLAND ROADS

In the winter of 2010 I was driving back from 'The Tip' of Cape York Peninsular, and I stopped at the Bramwell Junction Roadhouse for petrol. While at the roadhouse I asked a local man if he had heard rumours of Tasmanian Tigers living on Cape York Peninsular, as I have over the years. He replied in a very definite and convincing manner, 'They are here on the Cape. You should go to Portland Roads because there is a lady there who knows all about them living there!' He then said, 'When you arrive there, ask anyone there where she lives, because everyone there knows her, and it's a very small township.' I then told him that I was intending to visit Portland Roads on my way back anyway, so I'll have a chat to her.

When I arrived at Portland Roads I was told where she lived by a local man, and I then visited her at her house, which was in this hamlet. When I arrived I told her why I had come to see her. She was approximately sixty years old, and was friendly and happy to tell me what she knew. She told me the following about what she knew about Tasmanian Tigers:

'I sometimes heard loud unusual screams at night near my previous house, which was just outside of Portland Roads itself. These screams sounded like a woman being strangled, and I believe that these Screams were probably made by a Tasmania Tiger.' Note: During my research, I had found that Thylacines had what was known as a 'screaming woman call.' I did not tell her about my knowledge of this call.

She continued, 'My original house had a 500m-long straight dirt road running from it, then it had a 90 deg. bend in it, and then it ran a further 200m approximately where it joined the main dirt road into Portland Roads. One day, after overnight rain that had removed all tyre marks and animal tracks from it, I went for a walk along it and I saw very clearly only two sets of animal tracks in the soft damp dirt. They were Tasmanian Tiger footprint tracks that had come out of the vegetation near my house onto the road. These footprint tracks were side by side as they walked away from my house along the road.

One set was an adult's, and the other was those of it's baby. They had walked side by side for 100m to 200m , then the baby must have got tired, because I could see by their footprints that the adult had crouched down so that it's baby could climb into it's rear-opening pouch.

The only footprints after this were those of the adult, which continued along the road to the 90 deg. bend, where they disappeared into the vegetation, because the adult continued walking in the same direction as it had walked along the road.'

I then asked her if there was any evidence that the adult had eaten the baby one by her analysis of the footprints. She replied 'definitely not, because their clear footprints in the dirt showed that the baby had climbed into the adult's pouch.'

I then asked her, 'what was the size of the footprints?,' and she replied, 'The adults were as big as a large dog's and the baby's were this wide.' She then showed me the width of the baby's footprints by holding up her hand with a distance of approximately 25mm between her thumb and forefinger.

She also told me that there is a dam near her original house, and that she had also seen both Tasmanian Tiger, and very unusual large footprints, along with common native animal ones around the edge of the water in it. I then suggested that she set up a night-vision camera at this dam, so that she can see the animals that drink from it. I also suggested that she takes photographs of these footprints with a ruler beside them, to indicate their size. She also told me that one of her female relatives, who had also once lived at Portland Roads, had made plaster castes of strange animal footprints while living here.

Staying with her was her relative, who was approximately 20 years old, and he also told me that he also knows that Tasmanian Tigers live in the Portland Roads area.

This woman appeared to have a lot of knowledge about the native animals that live in the Portland Roads area. Both she and her relative appeared to be very honest people.

Figure 18 Looking north from the hamlet of Portland Roads, showing typical rain forest-covered mountains of the area. Author's collection.

Figure 19 The small beach at Portland Roads hamlet, taken from the WW2 jetty site. Author's collection.

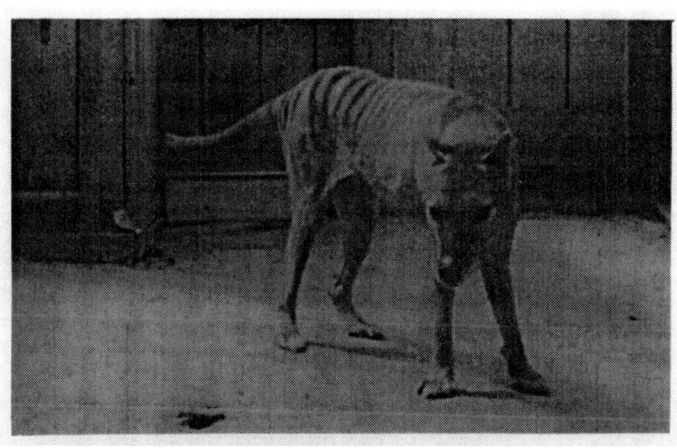

Tasmanian Archive and Heritage Office: ph30/1/5846

CAPE YORK PENINSULA

In May 2013 I spent three days in Cooktown, and while there I mentioned to a local man, who was approximately 45 years old, that the previous week I had returned from a holiday in Tasmania. I told him that during this holiday I met approximately 16 people who told me about their own Tasmanian Tiger sightings, or other sightings that they knew about since they officially became extinct.

To my surprise, because I had not told him about what I have been told about them living on Cape York Peninsular, the man then replied, 'There are Tasmanian Tigers in all parts of Cape York Peninsular also.' I then asked him in an astonished tone of voice, 'How do you know that!?' To which he replied, because National Park Rangers there, and people living there have told me. I then asked him, 'How long have you known this for?' To which he replied, '25 years, since I first came to live in Cooktown.' Note: I had not told him of my knowledge of them living on 'The Cape.'

He also told me that during the last Ice Age, Torres and Bass Straights did not exist, because Cape York Peninsula was joined to New Guinea, and Tasmania was joined to Victoria, enabling Tasmanian Tigers to walk between these three areas. Note: Scientists believe that this was due to the sea level then being approximately 130m lower than it is at present, because a great deal of ocean water was locked-up in the polar ice caps.

The next day I went to the National Parks Office in Cooktown, and told a Ranger there what the man had told me. I got the impression from the ranger that he knows that Thylacines live on Cape York Peninsular, but he is not allowed to tell me this. He did say however that they would live in the rugged mountain ranges on Capc York Peninsular. During our conversation he suddenly exclaimed, 'Thylacines have extremely sensitive hearing!' This sudden statement by him made me think that he knows a lot more about living Thylacines than he is willing to tell me about.

In June 2013, when I was again in Cooktown, I told a middle-aged woman, who lived there that 'over the years I have heard rumours of the 'extinct' Tasmanian Tiger living on Cape York Peninsula,' to which she replied, 'I know that they live on The Cape.' I believe that this woman was telling me the truth because of the 'matter of fact' way that she replied to me.

Also, during this same visit to Cooktown, I had a meal in the Country Women's Association, (CWA), room, and I mentioned to two local women who I met there, that 'I had heard rumours over the years that the 'extinct' Tasmanian Tiger is living on Cape York Peninsula.' Both of these 50-60 year old women then replied that they know that Tasmanian Tigers live on Cape York Peninsular.

I then asked them what they knew about them living on 'The Cape.' One of them replied that she had lived in the town of Bamaga for five years, and she, and other people who lived there knew that Tasmanian Tigers are living on 'The Cape.' It appeared to me that from her manner of answering, and what she told me, that their existence was common knowledge in Bamaga. <u>Note:</u> Bamaga is very close to 'The Tip' of Cape York Peninsula.

The other lady then replied that she knows that Tasmanian Tigers are in the Cooktown area, and that several years ago there was a newspaper article detailing the sighting of one at the town of Rossville. <u>Note:</u>

Rossville, which is a very small town is in an area that has large rainforest-covered mountains nearby. There is also a lot of rainforest in that general area. Rossville is approximately 39km from Cooktown, on the 'coastal road' that leads to Cairns.

I believe that both of these women were telling me the truth, because they told me what they knew about Thylacines in a very 'matter-of-fact' manner.

The following photographs, (Figures 20 to 44), show some of the locations, in sequence, as you travel north, on the road between Mount Malloy and 'The Tip' of Cape York:

Figure 20 View looking east from the road between Mount Carbine and Bob's Lookout. Author's collection.

Figure 21 View looking east from Bob's Lookout, on the road between Mount Carbine and the Palmer River Roadhouse.. Author's collection.

Figure 22 Lookout at the Byerstown Range on the road between the Palmer River roadhouse and Lakeland, which is approximately 12km away by road, view looking north east. Author's collection.

Figure 23 Several kilometres south of Laura, looking south back along the road, which is now sealed, towards Lakeland. Author's collection.

Figure 24 A creek crossing at Musgrave on the road to Cape York. Author's collection.

Figure 25 The Archer River Roadhouse on the southern bank of this river, in the early morning. Author's collection.

Figure 26 The Archer River, from the road bridge near the roadhouse. Author's collection.

Figure 27 The road between Archer River and Morton Telegraph Station. One of very few high places on the road where you can see the vast Cape York Peninsula forest stretching away to the horizon. Author's collection.

Figure 28 The road to Cape York, at the Charlotte Creek road crossing, 17km north of Archer River. Author's collection.

Figure 29 The old Moreton Telegraph Station entrance. Author's collection.

Figure 30 The Wenlock River from the road bridge, near the old Morton telegraph station. Author's collection.

Figure 31 A sign on the tree at the top of the photo, stating, 'we were here in a boat---.' South side of the Wenlock River road bridge, near the old Morton telegraph station. Note: the river behind, and approximately 4m below the ground level at the tree, and the metal safety barrier on the side of the road at the bottom right. Author's collection.

Figure 32 One of the extremely numerous large 'Cathedral' termite mounds found on Cape York Peninsula. On the road between old Morton telegraph station and Bramwell Junction roadhouse. Author's collection.

Figure 33 Tree at Bramwell Junction roadhouse with dislodged vehicle registration plates, due to the corrugated gravel road. Author's collection.

Figure 34 The turnoff to Captain Billy Landing between Bramwell Junction roadhouse and the Jardine River ferry. Author's collection.

Figure 35   Heathland National Park, view looking south from the road between the turnoff to Captain Billy Landing and the Jardine River ferry. Author's collection.

Figure 36 Palm trees and thick vegetation behind author's vehicle, between Heathland National Park and the Jardine River ferry. Author's collection.

Figure 37 The road between Bamaga and the Jardine River, showing the typical dense vegetation of 'The Cape.' Author's collection.

Figure 38 My 'morning tea' rest stop between the Jardine River and Heathland National Park,, taken looking south. Author's collection.

Figure 39 At my 'morning tea' rest stop shown in the previous photo, looking east from the edge of the road into the very thick vegetation. Author's collection.

Figure 40 The Jardine River, looking downstream from the vehicle ferry. It is approximately 48km south of Bamaga by road. Author's collection.

Figure 41 The road between the township of Seisia and 'The Tip' of Cape York. Author's collection.

Figure 42 The junction of the roads to Somerset and 'The Tip' of Cape York, showing the thick rainforest at this location. Author's collection.

Figure 43 The most northerly point on the Australian mainland, sign at 'The Tip' of Cape York, looking north to Eborac Island. Author's collection

Figure 44 The marker disc that is approximately 400 metres from 'The Tip' of Cape York, taken looking south west. Author's collection.

CROYDON AND ENVIRONS

In October 2011, I met an elderly woman who was a long-time resident of Croydon, which is in the southern part of Cape York Peninsula. Her very wrinkled and suntanned skin showed that she had spent a lot of time in the sun. I found her to be a very friendly, easy to talk to, and a 'down-to-earth' woman, who knows a lot about the bush in far north Queensland.

I told her that I had heard rumours of Tasmanian Tigers living on 'The Cape.' She then told me about a strange 'big dog' that she, and other people have seen, in the distance, over the years at Belmore Dam, which is on the outskirts of Croydon. She also seemed to know about Thylacines, but didn't tell me anything specific about what she knew.

After I had arrived back home, I phoned her, and asked if she could give me any more details about what she knows about Tasmanian Tigers. She then told me that when she was a young woman working on a property, that was less than 100km north of Croydon, there was also an old aboriginal lady there who would tell her that her family had 'Stripey Dogs' as pets, when she was a young child. The woman then told me that she believes that these would have been Tasmanian Tigers. I then told her that it is recorded that some people in Tasmania also kept them as pets.

Figure 45. The road between Normanton and Croydon, showing part of the vast Savannah forest of northern Australia. Author's collection.

SMITHFIELD AND TINAROO DAM

In the winter of 2011, when I was at Ravenshoe on the Atherton Tablelands in North Queensland, I met a man who was apparently fifty nine years old. I told him that in the last couple of days I had met a woman who told me that she, and her three friends, had seen a Tasmanian Tiger walk in front of their car while they were shooting rabbits at night.

He then replied, 'I saw a Tasmanian Tiger 50 years ago, when I was nine years old, and I can still very clearly remember it, as if I had seen it last night.' He then continued by saying, 'My parents then owned, and we lived on the land at Smithfield, on which the annex of the James Cook University is built. My father would quite often tell me and other family members when he arrived home from work, at about 9pm, that he had just seen, in the headlights of his car, the Tasmanian Tiger again walk across the Captain Cook Highway at a point that was a few kilometres south of its junction with Trinity Beach Road.'

Note: I have driven on this section of highway during the last three years, and I saw that the rainforest-covered mountains extend right down to the western edge of this highway in this area, so I assume that these sightings were probably seen adjacent to this rainforest area, and the Thylacine was probably living in the mountains in this area.

He then told me that 'My father would tell us that he sees it cross the highway in his car's headlights at the same place, and time in the evening, and one night I was in the car on this part of this highway as my father drove home, and the Tasmanian Tiger walked across the highway in front of our car, I saw it very clearly in the beams of the headlights. My father said to me as we watched it, take a good look at it, because it is a rare animal.'

This man also told me that Tasmanian Tigers had been seen on the Atherton Tablelands over the years. He also told me that, 'There is an old man who I know, who was originally from Eastern Europe, and he often walks in the rainforest at the Tinaroo Dam, on the Atherton Tablelands. He has told me that he has seen Tasmanian Tigers several times during these daytime walks.'

MOUNT BARTLE-FRERE

In 2011, I was having lunch at the lookout on the road between the Atherton Tablelands town of Herberton and the town of Millaa Millaa. This lookout has panoramic views to the coast, Millaa Millaa, and Mount Bartle-Frere, which is covered in rain forest, and is the highest mountain in Queensland at 1622m (5322 feet.)

While I was there, a twenty two-seat 'Eco' tourist bus arrived with overseas tourists in it. I told it's middle-aged driver about some of the Queensland Tasmanian Tiger sightings that I have been told about, thinking that he may also know about them, and he replied, 'Bushwalkers have also seen Tasmanian Tigers on Mount Bartle-Frere,' and then he pointed out this mountain to me. Note: It is approximately 35km from this lookout.

Figure 46 Mount Bartle-Frere in the distance, seen through the humidity haze, taken looking east from the lookout on the road between Herberton and Millaa Millaa. Author's collection.

EAST OF GEORGETOWN

One afternoon In early October 2011, I was driving for the first time on the isolated 170km-long gravel road that runs from just west of Mount Garnet to the small coastal town of Kennedy. This road passes through forests for almost it's entire length, and I had not seen any vehicles on this road until I had driven 79km along it towards Kennedy. I then saw a 4WD vehicle coming slowly down a grass-covered slope towards me from a derelict domestic garden. I stopped my vehicle and waited until the middle-aged couple in it arrived. I asked them about the road conditions further on, and about

Blencoe Falls, which I wanted to see, and they gave me the requested information.

I assumed that they were local land owners, but they later told me that they lived on the Atherton Tablelands approximately 100km to the north. Because this whole area where we met was thickly forested, I thought that this whole area would be a suitable habitat for Thylacines, and that this couple may know about them living here, so I told them that I have heard rumours of Tasmanian Tigers living on Cape York Peninsular, and both Captain Cook and Joseph Banks recorded in their 'Endeavour' voyage journals in 1770 that there were 'wolves' at what is now Cooktown. I then said, 'You have to ask the question, what Australian animal did the crew members of the 'Endeavour' see, to make them think that they had seen a wolf, even though Captain Cook also recorded that a dog often came to be fed in their camp, while they were at what is now Cooktown?'

To my surprise the woman replied, 'I saw a Tasmanian Tiger about twenty years ago!' and she then said, with utter relief in her voice, 'You are the first person who I have ever told this to, because I was scared that people would think that I was mad!' She repeated this twice more during our five-minute conversation, each time with the same utter relief in her voice. She really wanted me to know that I was the first person who she has told about her sighting. It was obvious to me that she had been 'bottling up' this sighting for the last twenty years, but now, finally, she had met a person who would not think that she was mad, or ridicule her.

After she had said this the third time, I said to her, 'Don't worry about what people think about you, have a look at the website Tassietiger.org, and do a Google search on Thylacine sightings, because you will see a great many Tasmanian Tiger sightings recorded on the numerous websites that result from this search.'

I then asked her where she had seen it, and she replied, 'It was about half way between here and Georgetown, at about midnight one night, when three of my friends and I were rabbit shooting from inside our parked car on a gravel road.' I then asked her for the details of their sighting, and she replied, 'As we sat in the car with its headlights shining down the road, it casually walked across the road a couple of yards in front of it, and then stopped and looked around for a while,' She then told me that 'after looking around for a while, it casually continued walking across the road in the same direction that it had been walking, and on into the bush until it disappeared.'

She then told me, 'The four of us sat there with our mouths wide open, because we all knew what it was, but we couldn't believe what we were looking at.' She also said to me, 'I wish that I had a camera that night, instead of a rifle.' I then asked her if it was worried about walking and stopping in the full glare of the headlights, and being near to the car, and she replied that it was not worried at all about the headlights, or being near to the car.

I felt really sorry for the woman because she had been too scared to tell anyone about her sighting, although she had obviously told her husband about it, because he sat quietly next to her in their vehicle, listening to the conversation that I was having with his wife.

I also suggested to them that they read about the wolves that are recorded in Captain Cook's and Joseph Banks' journals, edited by J.C.Beaglehole, which may be available as reference books in the Cains Library.

<u>Note:</u> From what she told me, their sighting would have been in the general area of the small town of Mount Surprise.

Figure 47 A road junction on the 170km-long road from Mount Garnet to Kennedy, showing typical forest of that area. Author's collection.

Figure 48 A properties sign at the northern end of the road from Mount Garnet to Kennedy, with typical thick bush of area. Author's collection.

Figure 49 Typical bushland in the area to the east of Georgetown, Queensland. Author's collection.

he believed that it must have been given this name because there were Tasmanian Tigers in that area, and he planned to search for them there.

Note: On the Royal Automobile Club of Queensland road map of the Cairns to Townsville area, there is a 'Tiger Hill' situated approximately 60 km NW of Cardwell. This is probably the 'Tiger Hill' that the man saw on the map.

Figure 52 Rugged rainforest-covered mountains in the distance near the coast of north Queensland. Author's collection.

Figure 53 Rugged rainforest-covered mountains near the coast of north Queensland, with a sugar cane farm in the foreground. Author's collection.

Figure 54 Rugged rainforest-covered mountains near the coast of north Queensland. Author's collection.

Figure 55 Rugged rainforest-covered mountains near the coast of north Queensland, with a banana farm in the foreground. Author's collection.

Figure 56 Rainforest-covered mountains behind the sugar mill at Tully, near the coast of north Queensland. Author's collection

Tasmanian Archive and Heritage Office: ph30/1/5847B

ROLLINGSTONE

In June 2003 an employee at James Cook University in Townsville, Queensland told me that a farmer living at Rollingstone, which is on the Bruce Highway between Townsville and Ingham, has informed this university that he has seen Thylacines at Rollingstone and in the nearby area over the years.

COLLINSVILLE

In 2009 I drove to the gold-mining town of Ravenswood in Queensland via the small town of Clare. When I arrived in Ravenswood, I asked directions from an elderly man who lives there. He appeared to be a very honest and polite man.

During our conversation he mentioned to me that he drives to Cape York Peninsular quite a lot, so I asked him if he had heard 'rumours' of Tasmanian Tigers on the cape, as I have, to which he replied, to my surprise, 'No, but I saw one once at Collinsville.' <u>Note:</u> Collinsville is to the east of Ravenswood.

I then asked him the details of his sighting, and he replied, 'It was approximately 150 yards away in the dry grass, and initially I thought that it was just another Dingo, but, what drew my attention to it was that as it started walking, it bobbed up and down, like this, and this is when I knew that it definitely was not a Dingo, because Dingos don't walk that way.' <u>Note:</u> As he told me that 'it bobbed up and down like this', he actually bobbed up and down approximately seven times by bending his knees.

He then told me that, 'This bobbing walk drew my attention to it, and made me I look at it very intently, and this is when I saw the black stripes across it's back, and I then knew that it was a Tasmanian Tiger. It was definitely not a Dingo, and it continued it's bobbing walk away from me until it disappeared in the distance.'

I then asked him at what part of Collinsville he had seen it, and he replied, 'It was on the east side of the town, at Pelican Creek.' Note: I had never heard of Pelican Creek, but when I returned home from this trip I used 'Google Earth' on my PC, and found that it was only approximately 500m east of the most easterly houses in Collinsville, and as this man had told me, I saw that it was a grassy area with few trees where he saw it.

In 2012 I met a middle aged man who told me a lot about his extensive knowledge of the native animals in Queensland. He also told me about his many bush-camping trips to isolated areas in the eastern parts of northern Queensland. This man appeared to me to be very honest, and very mature minded.

After he told me that he has camped in the Collinsville area, on the cattle property owned by one of his friends, I told him the details of the Collinsville sighting that I have recorded above. To which he replied that his friend has also seen Thylacines several times, and at several locations on his property over the years.

He then told me that I should camp overnight at a specific location in the Collinsville area, because, 'At night I may see an animal there.' He would not tell me that I would see a Thylacine there, but he again told me that I should camp there overnight. I believe that he was giving me a very strong hint that I would see a Thylacine there.

He told me that he has camped overnight at this location several times, and also, it is not on his friend's property. This location is approximately 6km west of a very extensive and rugged mountainous area. He also pointed out this location for me on my topographic map of the area.

Figure 57 Typical bushland in the Collinsville area. Author's collection.

Figure 58 Emus in the Collinsville area. Author's collection.

SPRINGBROOK

In November 2013 I met a local middle-aged man who appeared to me to have a great deal of knowledge about native animals, including snakes. He told me that in 1980, he and his mate were walking at night up a road that leads from a valley, which has an Aboriginal-sounding name, to the town of Springbrook, which is in the mountains to the west of the Gold Coast area of south east Queensland.

He told me that they were looking for nocturnal snakes on this road, when suddenly a Tasmanian Tiger walked out onto the road just in front of them, but as soon as it saw them it darted back into the bush.

He believes that because it was a perfectly calm night, it did not smell their body odour, because if it had, he believes that it would not have made an appearance, which lasted for only a few seconds.

He was a sensible man, who told me a lot about some of the native animals of Queensland. He told me that over the years he has walked and camped in a lot in the very isolated areas in Queensland

I believed that he was telling me the truth, but also because he told me to visit a particular small isolated area to find a certain animal, which I subsequently did, and I found that the animal was there.

Note: I believe that it is too much of a coincidence, and against the laws of probability, that I have unexpectedly heard so many times about Thylacines being seen in Queensland, if what I have been told was not true. I also believe that eventually someone will manage to take a movie of one with their mobile phone etc., and then be willing to disclose this to the public.

INFORMATION FROM AN ABORIGINAL MAN

In approximately 2000 I spoke to an Aboriginal man who was working at an aboriginal display for tourists in northern Queensland. I asked him about the display, and I also asked him if he knew of any aboriginal paintings or carvings of Thylacines in northern Queensland, because I had seen photographs of them in Western Australia and in the 'Top End' of Australia. He replied that he knew of two of them in northern Queensland.

Later during our conversation I told him that Thylacines were now extinct on the mainland of Australia. From then on during our conversation I got the distinct impression by his attitude towards me, and by his tone of voice, that he thought that I was completely ignorant of the fact that they are not extinct, i.e. I did not know what I was talking about.

INFORMATION FROM AN EX-QUEENSLAND NATIONAL PARKS RANGER

In 2013 I met an ex-Queensland National Parks Ranger, who, told me that in all honesty, nobody really knows what animals live in northern Australia, because, there are so few people there, and, because it is such a vast area, where very little research on its fauna has been carried out.

He told me this in reply to me telling him about some of the Thylacine sightings in Queensland that I have been told about.

INFORMATION FROM A QUEENSLAND NATIONAL PARKS OFFICER

In 2011 I was told by a National Parks officer, who was stationed in Queensland, that Thylacines were living in the area where the officer was stationed! The officer told me this in a matter-of-fact manner, after I had said that I have heard rumours of Tasmanian Tigers living on Cape York Peninsula, and that both Captain Cook and Joseph Banks had recorded in their journals in 1770, that there were ‘wolves’ living at what is now Cooktown.

THYLACINES EXCAVATE SOIL BANKS

In June 2013, I phoned an officer at the head office of the Department of the Environment in Queensland. I told the officer about some of the things that I have been told about Tasmanian Tigers being sighted in Queensland.

To my surprise, the officer then told me that Thylacines excavate a cave in soil banks, for use as a den to live in. I believe that this was the officer's indirect way of telling me that Thylacines are living in Queensland.

The officer also told me that the vast, very sparsely populated areas in Queensland have not been checked yet for the types of native animals that live in them. I got the impression that the officer has travelled to these vast areas.

Tasmanian Archive and Heritage Office: ph30/1/5844

Tasmanian Archive and Heritage Office: AA193/1/1003

CHAPTER 2 THYLACINES IN TASMANIA

PHOTO OF THYLACINES IN NORTHERN TASMANIA

In approximately 2000, on my return to Sydney on the 'Spirit of Tasmania' ferry from Devonport in Tasmania, I met a middle-aged couple, who I assumed were returning home to Sydney, as I was. The man then told me that they live in Tasmania and were going to Sydney for a holiday. I then replied that during the last few weeks in Tasmania, I have met several people who claim to have seen a Tasmanian Tiger, or know someone who has seen one. I was hoping that they may also have seen one, and will tell me about it.

The man replied, in 1996 an honest friend of his was sitting alone in his broken-down car on an isolated road at night in the north of Tasmania. He told me the name of the nearest town, which had an aboriginal name, and which I had never heard of before, but it may have been in the Mawbanna area. I can't remember because of its unusual name. The man told me that his friend walked from his car into the nearby bush to relieve himself, and a couple of minutes later as he returned to his car he could hear the sound of a large animal rustling the vegetation near his car. He crept towards the car as quietly as he could so that he did not scare the animal away, and he removed the camera that he had in the pocket of his jacket, because he wanted to photograph whatever it was. As he got closer to his car he thought that he could see what looked like two dogs standing next to the driver's side of his car, which made him think that they may be Thylacines. He took a flash photo of the animals, and as the flash operated he thought that he could see two adult Thylacines with two 'pups.' As soon as the flash operated the four Thylacines bolted away.

The man also told me that one of this man's friends was a professional photograph, and that the man who took the photograph didn't want it known by anyone, so he went to his friend and told him that he is almost certain that he has managed to take a photograph of two adult Tasmanian Tigers, and their two 'pups' standing next to his car. He asked him if he was willing swear an oath to him that he will not tell anyone about him developing and printing the photograph for him, because if he won't swear an oath he will take the film to another professional photograph who will swear an oath.

The professional photographer then said that he will swear the oath for him, which he did, and subsequently developed and printed the film for him.

He told me that the film shows two adult Thylacines with their two 'pups' standing next to the car. He was very keen to tell me that it is a genuine photograph that has not been faked.

ROAD KILL ON THE ROAD TO QUEENSTOWN

In approximately 2000, while I was on a holiday in Tasmania, I met a serious, and apparently well educated, man, who was approximately 35 years old. He told me that early one night, several years before, he was driving alone in his car along the very mountainous and isolated road that leads to Queenstown, on the west coast, when he saw a 'strange' road kill on the left side of the road.

He stopped and reversed his car to the dead animal, and found that it was a freshly killed, very young Tasmanian Tiger that had been hit by a vehicle.

He took it home with him because it was definitely a Tasmanian Tiger, and because he planned to take it to a National Parks office, to show the officers there that they were still alive. He took it to the office the next day and left it with an officer, who told him that 'It is a Quoll,' but the man knew what Quolls look like, so he disputed the officer's identification because it was obviously a Thylacine. The officer then told the man that he will send it away so that identification tests can be carried out on it. The officer asked him to return in a week or so for the result of the tests. He did this, and the same officer told him that the tests showed that 'it was a Quoll.' The man then asked the officer for it's body, the officer replied, 'it has been thrown out.'

The man was certain that this animal was a Thylacine, and he was very sorry that he had given it to the National Parks officer. He told me that if he known that the officer was not going to return it to him, he would never have given it to him.

I then told the man that if I had found it, I would have taken dozens of photographs of all parts of it's body, and teeth etc., with a scale ruler in each photograph. Then I would have preserved it in a plastic bag in the freezer of my fridge, before taking it to the National Parks office.

I also told him that I would not have told the National Parks staff that I had taken the photographs of it, in case the officers claimed that it was not a Thylacine, and refused to give it back to me. The man replied, 'I wish that I had done that, because I never thought that they would tell me that it was not a Tasmanian Tiger, and not return it to me, because it obviously was one.

WEST OF HOBART

In 2003 I was on holidays in Tasmania, and I visited the recently-opened Axeman's Hall of Fame building in Latrobe, where there was a retired Timber Cutter working as the guide and Information Officer.

I mentioned to him that I had met people living in country towns in Tasmania, who are adamant that they, a relative, a work colleague, a neighbour, or a friend had seen a Tasmanian Tiger. I then asked him if he, or his workmates had seen any when they were timber cutting He replied, 'No, but I know a man who has heard their calls at night when he camps in The Great Western Tiers mountains.' He then told me that this camper 'spends two or three weeks at a time camping there at all times of the year, even when it snows. He is a real nature lover, who knows all of the calls of Tasmanian animals, and, he believes that the loud howling-like calls that he hears in the distance there can only be made by Tasmanian Tigers.'

The next day I drove to a small town to the West of Hobart, and booked a room for the night. That evening I told the manager there about what the man at the Axeman's Hall of Fame had told me the previous day. He replied, 'There is a man living in a house on the road that runs out to the Gordon Dam from here, and he knows all about Tasmania Tigers, you should have a chat to him about what he knows.' I then told him that I was planning to drive to the Gordon Dam tomorrow, so I will call in and have chat to him. He then gave me the details of how I can find this house.

The next day I called in to have chat to this elderly man, who told me that he was now a farmer, but he had been a Timber Cutter prior to retiring. I asked him if he, or his workmates had seen any Tasmanian Tigers when they were timber cutting. He replied, 'No, because we made too much noise.'

I then asked him if he knew anything about Tasmanian Tigers, as the manager had told me. He then told me the following about them:

'I used to quietly climb the forested hills behind my farm to watch one sleeping in daytime, in a rock shelter. It was several miles away from my house, but it is no longer living there.'

'Their droppings are approx. 50mm diameter and full of hair.'

'They have a stiff spine, not like a dog that can bend it's back into a 'U' shape, so, if they want to attack you by biting the back of your legs, they have to come at your back in a curved walk or run about two or three metres from you, and then come in from directly behind you.'

I asked him how often are Tasmanian Tigers now seen by himself, and by the other farmers in this general area? He replied, 'In the last two years the farmers had seen six of them.'

I then said to him that he and the other farmers should report all of these sightings to the National Parks Department or other government departments, but he replied, 'That is the last thing that we would do, because, we are scared that the government will force us off of our properties, pay us much less than they are worth, and then turn them into a National Park, or a Thylacine Reserve. He then told me that the farming families have lived on their properties for generations, and we don't want to leave this area.' He also told me, 'The Thylacines sometimes kill a few of our sheep, but we don't worry about it at all, because we are glad that they are still alive in our area.'

He also told me that he quite often hears one 'yapping' as it passes through the paddock next to his house in the early hours of mornings, and his wife is often asleep when this happens, so she does not hear them. He then told me that 'one morning at approximately 2am, the yapping occurred just outside their bedroom window, and his wife was also awake, so she also heard it. I told her that as soon as it was daybreak he will go into this paddock to see if he can see evidence of it. I did this, but saw nothing in the way of it's tracks, but as I was at the roadside front fence of this paddock, I looked at the 'ring-lock' fence on the opposite side of the road, and wedged into it, half way up it, was a freshly killed wallaby.'

He then told me that it was definitely not a 'dog kill' because of the way it had been killed and partly eaten. He told me that the yapping Tasmanian Tiger that they had heard several hours before was probably the one that he would watch in it's shelter, and also the one that killed this wallaby. He thought this because it looked like it had tried to drag the wallaby over the fence onto the road, but the fence was too high for it to drag it over it, so it tried to drag it through it.

He told me that it probably intended to drag it over the road, then over his paddock fence, through his paddock and then up into the forested hills behind his house, or up into it's shelter. He also told me that the dead wallaby was not wedged in the fence at dusk the day before he found it.

I then asked him what the yapping sound was like, and did it ever make other type of sounds, He replied that the nearest sound that he can liken it to was that of an Alsatian puppy yapping, and he had never heard them make other types of sounds.

Note: His wife was standing next to us during our conversation, and she did not dispute anything that her husband told me. I believe that this man was telling me the truth.

TEMMA

In June 2004, I met a couple who were approximately sixty years old, from the general area of where I live in Queensland. During our conversation I mentioned to them that I sometimes visit my relatives who live in Tasmania. The woman then told me that she originally came from the Don River area of northern Tasmania.

I thought it would be a good idea to tell her about some of the things that people in Tasmania have told me about their Thylacine sightings, thinking that she may also have a similar sighting that she would be willing to tell me about, and she did. The following is what she told me:

When she was at school in the early 1960s, over a period of approximately three years, she, her cousins, siblings, fiends, and their parents, would spend some of their school holidays, and holiday long weekends at her uncle's fishing hut at an isolated place named Temma, on the north west coast of Tasmania. She told me that there were only about a dozen holiday fishing shacks there, and that nobody lived there permanently.

She told me that they would catch Lobsters from the nearby ocean rocks each day, and cook them in the afternoons on an open fire in the middle of their un-fenced 'back garden', leaving the scraps next to the fire. Then, each evening at dusk, while there was still good visibility, they would watch quietly, from inside their shack, a group of approximately seven Tasmanian Tigers walk from out of the nearby sand hills, which were on the eastern side of the shacks, to the lobster scraps at the side of their fire.

I then asked her about their physical details, and how they behaved. She replied, that on arrival they would start eating the scraps next to the extinguished fire. They were an orange-brown colour with black stripes across their backs. There were, big fully grown, half size, and small ones. They were apparently a family group that never made any sounds. They had an extremely good sense of smell, because the westerly winds would carry their cooking smell inland to the forest where they lived. This forest was about half a mile away.

She also told me that after they had been eating happily for several minutes, Tasmanian Devils would arrive to also eat the scraps, while the Tasmanian Tigers were still eating. There was no aggression from any animal at this stage, all happily eating together, but, after a minute or two the Devils would start to fight each other for the food. This caused the Tasmanian Tigers to walk backwards away from the Devils, while all the time intently watching them fighting to make sure that they themselves would not be attacked by them. The Tasmanian Tigers would then walk back over the sand hills towards the forest, walking at the same speed as a man walking briskly. They would not return again until the next evening, when, exactly the same thing would occur. They were a very cautious and shy animal that did not want to be involved in any fighting. She also told me that that, as children, they took the Tasmanian Tigers there for granted, because they saw them there so often.

She told me that a joke that the children had among themselves at night, was to scare any child who was going to use the outside toilet, which was beyond the fire from their shack. They would say to the child, 'Watch out for the Tigers and Devils, because they are there. Make sure that you take the torch with you so that you can see them, and make sure that they don't attack you!'

She also told me that at this time all of the adults in her family, and their adult friends, would refer to the Thylacines as Hyenas.

About one year after the woman had told me this, I went on a holiday to Tasmania. I decided to drive to Temma to see if it was the same as she had described it to me. It was exactly the same, as I thought it would be, because, I believed that she had told me the truth. I saw that there was a long wire property-line fence separating the shacks and road from the sand hills, but recently I asked this woman if it was there in the early 1960s, and she replied that it was not there then.

Tasmanian Archive and Heritage Office: NS463/1/5/2

WYNYARD AREA

In approximately 2007 I met a man, who was approx. 25 years old, at Temma, which is on the North West coast of Tasmania. I told him that I had met Tasmanians who have assured me that Tasmanian Tigers had been seen by themselves, relatives, or friends etc.

I also told him that I had come to Temma to see if the terrain etc. was as it was described to me by a woman, who told me that in the early 1960s she saw Tasmanian Tigers outside the beach hut that she stayed in here. I also told him that I believed what she had told me, but I always wanted to come here to confirm that it is the same as she described to me, and it is.

To this he replied that when he was a teenager he, his friends, and his child relatives were walking down a country road in the Wynyard area of Tasmania in the daytime, when suddenly approximately thirty metres ahead of them a Tasmanian Tiger walked out of the bush onto the road. He told me that it was plainly visible to all of them, and that they could see the black stripes across it's back. It was visible for approximately one minute allowing them all to have a good look at it as it walked away from them down the road, before walking into the bush on the side of the road.

He then told me that they were keen to tell their parents about what they had seen, but they knew that their parents would say to them that they were making up a story, so, he and the other older members of the group discussed how they could convince their parents that they were telling the truth. They decided that when they got home they would ask the man's six year old female cousin to draw a picture of what she had seen, if the parents did not believe them.

The older members of the group made sure that she did not hear what they planned for her to do if the parents did not believe them. They came up with this plan because they knew that their parents would believe the six year old.

When they got home they found that their parents did not believe that they had seen a Tasmanian Tiger, so the older members of the group asked the six year old to draw a picture of the animal that she had seen. She then drew a picture of a Tasmanian Tiger, including the stripes across it's back. This convinced the parents that the children were telling them the truth.

BEECHFORD

In May 2012 a middle-aged local man who I met in a shop in the main street of Latrobe, told me that his brother plainly saw a Tasmanian Tiger, which was as big as a large dog, as he walked along the beach, in the daytime, at Beechford, on the north coast of Tasmania.

He told me that his brother saw it in 2000, and he could plainly see the stripes across it's back as it stood looking out over the ocean from the edge of the trees at the back of the beach. He told me that when it turned it's head and saw his brother approaching, it walked back into the trees and disappeared.

Because Thylacines apparently have an extremely good sense of smell, I wonder if this Thylacine had smelt odours that had been blown across Bass Straight from mainland Australia, and had decided to investigate their source, but was blocked by the ocean. It is known that the smell of springtime flowers and bushfire smoke can be smelt at sea hundreds of kilometres from the land.

Figure 59 A Beach at Beechford, looking east. Author's collection.

SIGHTINGS I WAS TOLD ABOUT DURING A HOLIDAY:

NEAR WARATAH

In May 2013 I went to Tasmania for a holiday, and within 1.5 hours of arriving at Launceston airport I arrived, in my hire car at Deloraine, where I met two women working in a shop. I mentioned to them that on previous Tasmanian holidays I had met people who told me about Tasmanian Tigers that they, a relative, colleagues, of friends had seen.

One of the women replied, 'one night, twenty years ago, my father who is very honest, and never lies to us, told me when he arrived home from Savage River, that he plainly saw a Tasmanian Tiger run across the road 20 to 30 metres in front of his truck when he was near the town of Waratah. He could plainly see the black stripes across it's back, and there was no mistaking what it was.'

I told a middle-aged woman customer who entered this shop several minutes later, what the woman had told me, about her father's sighting, and she replied that two years ago, a very honest lady friend, of her own age, had seen one very clearly, including it's stripes, cross the road 20 m in front of her car as she drove at night. I asked her where her friend saw it and she replied that she had told her, but she couldn't remember where it was.

I was amazed at how lucky I was to hear about these two sightings, in a shop, so soon after arriving in Tasmania. I thought that during this holiday, this will be a very good way of initiating a conversation with strangers in the country towns, to see if they would tell me about Thylacines that they may have seen, or know about, so, I would initiate a conversation by saying to them the following:

'You would not believe it, but I had only been in Tasmania for about one and a half hours, when I met two women in a shop in Deloraine. One told me that her honest father saw a Tasmanian Tiger twenty years ago near Waratah, and the other woman told me that she has a very honest lady friend who saw a Tasmanian Tiger only two years ago!' I would then tell them the details of my conversation with both of them, and then wait for their response.

Approximately twenty three people subsequently readily told me the details of their own Thylacine experience, or knowledge, (i.e. almost everyone I told about what these two women had told me!), From what they told me, I got the distinct impression that Thylacines are well and truly alive in the mountainous and forested areas. The following are some of the things that I was told by using this technique on this holiday:

Figure 60 Mountains at Waratah. Author's collection.

2. SOUTH OF SCOTTSDALE

One middle-aged man who I met at Scottsdale told me that an old ex Bullock Driver, who lived in an isolated house in the forest-covered mountains to the South of Scottsdale, had regularly fed Thylacines at his house. He also told me that they have a very acute senses of smell and hearing, because if a visitor came to his house, the Thylacines would not go there to be fed for three or four days, because they could smell the different odour of the visitor. The man who told me this pointed to the tree-covered mountains in the distant south during our conversation, saying that 'that is where the old man lived.'

3. THYLACINES AVOID FIREARMS

Another man told me that Thylacines have learnt that if they smell gunpowder smoke from firearms, then they are in danger, and because of this they have retreated to live in the mountains. He also told me that they are definitely still living in Tasmania.

4. TASMANIAN MUSEUM AND ART GALLERY

I told a woman, who works at the Tasmanian Museum and Art Gallery, what a man on Cape York Peninsular had told me in 1993, about him seeing a Thylacine three times as he camped with his dog in darkness one evening at Iron Range, on Cape York Peninsula. She then replied, 'That is typical of the Thylacine sighting reports that this museum receives from people who see them as they are camping at night in Tasmania.'

5. TOMAHAWK

When I was at the small town of Tomahawk, which is on the north east coast, I met a local man who was walking briskly, for exercise. As he passed me I initiated a 'Thylacine conversation,' and he replied, 'I saw one in northern Tasmania,' but when I asked him for the details of his sighting, he refused to give me any more information.

6. BEULAH

A middle-aged customer who I met in a supermarket in Sheffield, told me that when he was in primary school in 1967 or 1968, his father, in mid-afternoon, swerved his truck off of a road to miss a Tasmanian Tiger near the small town of Beulah, which is to the south east of Sheffield. His truck nearly rolled over as it ended up in a deep roadside drain.

He also told me that this sighting appeared in a newspaper several days later, so the children at his school then started to make fun of him, by saying that his father was drunk, which he wasn't, when he drove into the drain, and his Thylacine sighting story was only an excuse for the accident. One girl in his class was particularly vocal about his father being drunk at the time, but two weeks later this girl's father also saw a Tasmanian Tiger in that area, so he took great delight in frequently telling her that her father had been drunk at the time of his sighting.

7. WILMOT

I learnt from a man at Wilmot that a lot of people in rural Tasmania shoot Wallabies and Kangaroos for food for their dogs, and some also shoot them for their own food, even though they know that they are 'protected' animals. When this man told me this I asked him if he thought that they would also shoot a Thylacine if they saw one, he replied, 'they would.'

8. A PROPERTY ON THE NORTH EAST COAST

I also met a middle-aged man who told me that an ancestor of his had a pet Tasmanian Tiger on his farm, which was on the north-east coast of Tasmania. He told me that the house on this farm was about five miles, (eight km), along a gravel road from the entrance gate, but if a visitor opened this gate, the Tasmanian Tiger would always 'react', letting his ancestor know that someone had opened the gate, and then always that person would arrive at the house in the time that it took to travel from the gate to the house. He told me that the Tasmanian Tiger had extremely good hearing and sense of smell. I believe that this man was telling me the truth.

Note: It is recorded that people did keep Thylacines as pets. I have no idea how the Thylacine could know that the gate was being opened eight km away if the wind was blowing odours and sounds from the house towards the gate. Maybe the warning calls of birds or animals near the gate were being communicated to other birds and animals closer to the house, and so on, until birds and animals near the house also communicated the warning calls, which the Thylacine heard and understood.

It is possible that the gate had squeaky hinges, a rattling metal chain, or a noisy latch, the sounds of which were carried on the wind to the house, where the Thylacine with it's extremely acute sense of hearing, would then 'react.' It would be an interesting exercise to find out just how sensitive the Thylacine senses of hearing and smell are by carrying out scientific tests on them.

9. SOUTH OF SMITHTON

I met an elderly local man in a cafe at Smithton, which is in the north west of Tasmania. He told me that his father had snared and trapped Tasmanian Tigers in the area to the south of Smithton, but he himself had not seen any.

10. GREAT WESTERN TIERS

A middle-aged man told me that Thylacines are known to exist in, and around the Great Western Tiers mountains in northern Tasmania.

Figure 61 The Great Western Tiers area in northern Tasmania. Author's collection.

I believe that all of the people mentioned above were telling me the truth when they told me about what they knew about Thylacines during this Tasmanian holiday.

A PAIR OF BREEDING THYLACINES!

In February 2014 I saw a man sitting in his car, which had Tasmanian Number Plates. I had seen this car parked in a local town during the last year or two, but never the owner of it. So, I decided to tell him about some of the Thylacine sightings that I have been told about in Queensland and Tasmania, hoping that he may be willing to tell me about sightings in Tasmania that he may know about. He was interested in what I was telling him, and he told me that he occasionally visits his family and friends in Tasmania. This man was approximately 50 years old and down-to-earth, straightforward, and I believe him to be an honest man.

He then told me that when his mother knew that she was about to die, she disclosed to him, and other family members that she had once seen a Thylacine at Golconda, where she lived in north east Tasmania. She also told them that she had never told anyone about her sighting, because she was scared of being ridiculed by people, but now that she was dying she felt that she could tell her family. He also told me that his mother's side of his family had known a lot about Thylacines.

I felt very sorry for his mother being so scared of disclosing her Thylacine sighting, even to any of her family members, until she was dying. I told the man that I was sorry to hear that his mother had died.

He then told me that he has family and friends in Tasmania, who keep him up-to-date about Thylacine sightings there. He also said 'have a look at this,' and then he showed me a movie, which he had on his mobile phone. It was a movie of a Thylacine walking around in a zoo cage.

He then really surprised me by saying, 'I know where there is a breeding pair of Thylacines in Tasmania, but there is no way I will tell you, or anyone else, where they are because I believe that their DNA etc. should be analysed by scientists, because they are rare, and I do not want them disturbed in any way.' I replied to him. 'I respect your wishes not to tell me their location, and I will not try to get you to tell me where they are.'

I then asked him if he would be willing to tell me if they are in the north, south, east, or west of Tasmania. He replied that they are in the north east of the state.

A week or so after meeting this man, I met a local woman who knows him, so although I believed what this man had told me, I thought that this would be a good opportunity to verify that he was honest, so I asked the lady if he was, and she replied, 'he is honest.'

I April 2014 I again met this man, when he excitedly walked up to me in a local town and told me that a mate of his in Tasmania had very recently told him that the breeding pair of Thylacines in north east Tasmania now have a pair of pups. I could see that he was happy about the pups being born, and so was I, because I believe that again he had told me the truth.

I then asked him how many years ago did his mother see the Thylacine at Golconda, and he replied, 'fifteen.'

In May 2014 I again 'ran into' this man, and he told me that a number of years ago he had a mate who was a builder. He told me that this mate lived in a house that was close to a creek at the end of Orchid Road, which is adjacent to the Conway Range National Park, in the Airlie Beach area, on the central coast of Queensland.

He told me that this mate had two large fearless pig-hunting dogs, because he had done a lot of pig hunting over the years, and would take these dogs with him on his hunting trips. He told me that very late one night, these two dogs started barking frantically outside his house, so he went outside to see what was causing this. He told me that when he got outside both dogs continued barking in a terrified manner and where staying at his mate's feet for protection, while they were looking into the rainforest in the direction of the creek, and that their behaviour was totally out of character for them.

He told me that his mate could not see what was causing his dogs to be so terrified, so he got his very powerful spotlight to see if he could see what was causing them to behave in this way. He told me that he saw in the beam of the spotlight, on the slope on the other side of the creek, a large dog-like animal in the vegetation, and that this animal's eyes reflected a colour that he had never seen before during his many years of hunting in the Queensland bush. He told me that his mate told him that he had no idea what this animal could be, so he replied to his mate that it would have been a Tasmanian Tiger, because they are known to terrify large fearless dogs like his. He told me that he is sure that his mate saw a Tasmanian Tiger that night.

He also told me that his builder mate had told him that he had seen all kinds of native animals during his many years of hunting, and that he was familiar with all of the colours of their eye shine, but he had never seen any native animal with the colour of the eye shine that he had seen that night.

I believed what he had told me about what his pig-hunting mate had told him, but In June 2014 I had an opportunity to possibly obtain more information about what the pig hunter had seen, and to verify what I was told was true. This was because I met a builder who had lived near Airlie Beach for many years. I assumed that the two builders could have met socially, as builders sometimes do. I asked him if he had ever known a builder, who had lived at the end of Orchid Road, near Airlie Beach, and who had done a lot of pig hunting with his two large fearless pig-hunting dogs. He replied that he had known this builder who lived there, and that he did have two large fearless 'pig dogs' which he took with him on pig-hunting trips. I then asked him if he had heard anything about his dogs being terrified by a large dog-like animal one night at their house, but he replied that he had not heard about that.

ANOTHER CONWAY RANGE NATIONAL PARK THYLACINE?

In July 2009 I met a man who told me that originally he had lived in Tasmania, so I told him about some of the Queensland Thylacine sightings that I had been told about, hoping that he may tell me about Thylacine sightings that he may know about in Tasmania. He was approximately thirty five years old, and he replied that his parents were also originally from Tasmania, but now they live at the eastern end of Monahan Road adjacent to the Conway Range National Park, near Conway Beach in Queensland. He then told me that his parents have seen large dog-like footprints on their property, and they believe that they are Thylacine footprints. He then told me that his parents, and their neighbours have also seen large 'strange looking' dog-like animals in the area of their properties, and his parents believe that these animals are Thylacines.

It is interesting to note that in approximate 2008, I met another man who told me that he has friends who live in

Monahan Road near Conway Beach, and that they at times have seen a large 'strange-looking' dog in the distance, at the edge of the rainforest, near their property. He told me this after I had told him about some of the Thylacine sightings I had been told about elsewhere in Queensland.

WEST COAST THYLACINES

In late 2014 I was told by a mature-minded, middle-aged Tasmanian man, who was holidaying in Queensland, that he had been told by outdoor workers on the west coast of Tasmania that they have seen a lot of Thylacines while working there over the years, but they do not tell people about their sightings.

He told me this, in a mater-of-fact manner after interestedly listening to what I had told him about some of the Thylacine sightings in Queensland that I have been told about.

CHAPTER 3 THYLACINES IN NEW SOUTH WALES

EAST OF TENTERFIELD

In 2011 I met, for the first time, a sensible young man who was in his early twenties. He told me that he had lived in Queensland for only six months after leaving his parents, 2 000 acre cattle property, which was to the east of Tenterfield, on the northern ranges in NSW. So I told him about another cattle property owner in that same area, who I had met a year or so before, and how she told me that she, and other cattle property owners in that area, had their cattle attacked at night by apparently some sort of 'Big Cat,' which left long and huge claw scratches down the sides of their cattle. He told me that his parents' cattle had also been similarly attacked at night.

He then suddenly, told me that when he lived on their property his father was driving alone in his vehicle on a dirt road near their property in the daytime several year ago, when, he saw a sandy-coloured 'dog' standing on the right-hand edge of the road approximately 100-200m ahead of him. The animal was not moving, so his father slowed the vehicle as he approached it, and stopped only a few metres from it so that he could have a good look at it, as it continued standing there.

His father told him that it was a Tasmanian Tiger, because it had black stripes across it's back, and he could see that it had a rear-opening pouch. His father told him that after watching it from his stationary vehicle for approximately one minute, it walked casually off the road and then through the nearby fence into the adjacent paddock until it disappeared in the distance. The man told me that just to the east of their property there was a thickly-forested mountain range, where he believed it lived.

NORTH COAST

In August 2013 I met a man who was approximately 45 years old, and he told me that he has worked at the Coffs Harbour Zoo, which is on the north coast of NSW. This man seemed to be a sensible, mature-minded and an honest man, and I have no reason to believe that he was lying to me.

He told me that when he worked there, at intervals of approximately once a year, a person, or persons would come to this zoo to report that they are sure that they had seen a Tasmanian Tiger in the rainforest-covered mountains to the west of Coffs Harbour.

He told me this after I told him about some of the Thylacine sightings in Queensland that I have been told about.

NORTH WEST OF BOURKE

In September 2014 I met a couple and their young children, who were on their way back home to NSW in their 4WD vehicle, after driving to Cape York Peninsular. I told the man, who appeared to be ‘well educated,’ about some of the Tasmanian Tiger sightings that I have been told about on ‘The Cape’, and elsewhere. He told me that over the years he has frequently gone, and still goes on feral pig and feral goat-shooting trips in the general area to the north west of Bourke in NSW. He also told me that his wife originally came from Bourke.

To my surprise he suddenly told me that three or four years ago while on a daytime hunting trip in that area, he saw a Thylacine running through the bush, killing goats, and then he watched it eating out six of their hearts and some of their other soft tissue.

He also told me that he followed it, and between several of the kills he saw it stop running after the goats, and then stand up on it's back legs, then it would resume it's chase. I assume that it did this to get a better view of the goats. He told me that it ran fast during each attack on a goat, and he felt their bodies, and verified that they were still warm, and therefore freshly killed by it. This infers that he did not see ever kill, but had to track it, until he came across it again eating their hearts.

Note: I had not told him that I knew that it has been recorded that Thylacines stand up on their back legs, and that they eat the soft tissue of animals.

He also told me that in this area he had once met an old aboriginal man, who told him that he had once been making a boomerang in the bush, and as he sat there making it, he had a sense that something was behind him, so he looked around and saw a Tasmanian Tiger just behind him, so he stood up. This caused it to run away. He told me that the man believed that it was about to attack him as he sat there.

It is interesting to note that a pack of Dingos have been known to follow a person who is walking in the bush, for a very long way, waiting for the person to sit, or lay down, causing them to think that the person is sick or injured. They then attack the person. Maybe Thylacines and Dingos have learnt to do this from each other, or it may be their instinct.

Note: Since I was told about this sighting, I have spoken to a woman who originally came from South Africa. I told her about how Dingos and Thylacines were known to follow a person for a long way through the bush, and she replied that Hyenas also do this. Maybe this behaviour by Thylacines is one of the reasons why they were once known as Hyenas.

He asked me for my name and phone number, which I gave to him, so that he can tell me about any other Thylacine sightings he has, or gets, from people in that area, because, he told me, people living there also now about Thylacines living there. Hopefully he will contact me in due course with more Thylacine information.

I believe that this man was telling me the truth. He was about forty years old, and he was interested in what I told him about the Thylacines sightings that I have been told about.

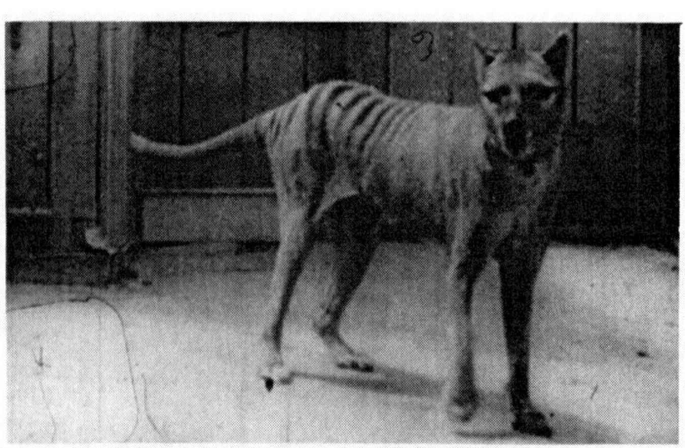

Tasmanian Archive and Heritage Office: ph30/1/5845

CHAPTER 4 THYLACINES IN VICTORIA

PHILLIP ISLAND AREA

In 2004 I met a young woman who told me that she had been taught at Phillip Island by a man who was the 'Tasmanian Tiger expert' there. She told me that he had told her that they kill sheep in this area by biting their spines at the back of their neck.

NINETY MILE BEACH

In November 2012 I was on a driving holiday in Victoria, and I made a point of visiting, for the first time, the small towns of Golden Beach and Loch Sport at Ninety Mile beach, because I had found the website 'tassietiger.org' that detailed Murray McAlister's search for Thylacines in these towns, and in the nearby areas. I wanted to know what the local residents knew about Thylacines in this area.

When I was at Golden Beach I asked several local residents if they had seen Thylacines here, and the three or four who I asked were adamant that they were in the area. A man who I met in a shop there told me that they were there, and that quite a few residents there had also seen them. He also told me that 'there had been a government-funded Scientist looking for them here, and this scientist had told him that he had found evidence of them in this area.' I believe that all of the people who had told me that Thylacines were in this area had told me the truth.

The next day I drove to Loch Sport, where I met a middle-aged man outside his house, which was on the north side of the road that runs from Golden Beach into this town. He told me in a very sincere manner that two years ago he saw a Thylacine just behind the back fence of his garden. He told me that he looked over this fence during the daytime, and clearly saw below him a Tasmanian Tiger that was only two or three metres away. I asked him how big it was, and he held up his hands to show me that it was approximately sixty five centimetres from it's nose to the end of it's body. He then told me that the Thylacine walked off into the bush that was behind his back fence. I also believe that he was telling me the truth. He also told me that quite a few residents of Loch Sport had seen Tasmanian Tigers in this town, or in the nearby area.

Figure 62 The reserve at the back of the beach at Golden Beach, showing some the thick coastal scrub of the area. Author's collection.

Figure 63 Ninety Mile Beach at Golden Beach, on the south coast of Victoria. Author's collection.

Figure 64 Lake at Loch Sport on the south coast of Victoria, taken looking inland. Author's collection.

WILSONS PROMONTORY

Also in November, 2012, I stayed at the 'Bimbi' camping ground near Cape Otway, and was told by a middle-aged local man there, that Tasmanian Tigers have been seen in the Otway Ranges in Victory. He also told me that there was an ex-lighthouse keeper who saw a Thylacine at the Wilsons Promontory lighthouse when he was stationed there.

An hour or two later I climbed to the top of the Cape Otway lighthouse, which was open to the public. At the top of it there was an elderly man who was explaining to visitors the technical details of the lighthouse. I told him about what the man at the camping ground had told me about the Thylacine sighting by the ex-lighthouse keeper. He really surprised me when he replied, 'that was me!' I asked him if this sighting was a hoax, and he replied that it was no hoax story, because he definitely did see it. I then asked him when he saw it, and he replied, 'it was in approx. 1984.' This man told me that he was originally from Tasmania, and that he had been stationed at all of the lighthouses on the north coast of Tasmania, and also at some of them on the Bass Straight coast of Victoria, during his many years of lighthouses keeping. He also told me that the story of his sighting was in the newspapers several days after he saw it.

Note: The man who told me that the ex-lighthouse keeper saw a Thylacine, did not tell me where this man was now living, or working.

Figure 65 West coast of Wilsons Promontory, taken looking south, and showing some of the thick coastal scrubland there. Author's collection.

Tasmanian Archive and Heritage Office: AA193/1/1002

CONCLUSION

After my conversations with the numerous people who have willingly told me about their Thylacine sighting knowledge, I have no doubt whatsoever that the Thylacine is not extinct on the mainland of Australia, or in the State of Tasmania.

Of course, a scientist will rightly say that the information that I have obtained from the people who told me what they know about living Thylacines is not scientific proof. The scientist will always want that indisputable proof of a live animal, because that is the only correct method that scientists will use to prove that an animal exists.

That being said, I do not believe that I, or anyone else has to believe that the people who tell you what they know about living Thylacines are liars.

Over the years I have found that there were only two people, who I believed were lying to me during my many conversations with people about what they know about living Thylacines. The information that those two people gave to me is not recorded in this book.

I will continue to try to find out what people know about living Thylacines.

SOME FURTHER READING

The following publications have information about Thylacines, and they are worth reading. Some of them may be difficult to obtain because they are out of print, but they may be available in libraries on inter-library loan, or from second-hand bookshops:

‘Tasmanian Tiger a Lesson to be Learnt’, by Eric Guiler and Philippe Godard.

‘The Last Tasmanian Tiger The History and Extinction of the Thylacine’, by Robert Paddle.

‘Thylacine The Tragic Tale of the Tasmanian Tiger’, by David Owen.

‘Shadow of the Thylacine One Man’s Epic Search for the Tasmanian Tiger’, by Col Bailey.

‘Cape York Peninsula A History of Unlauded Heroes 1845-2003’, by Lennie Wallace.

‘London Zoo from old photographs 1852-1914’, by John Edwards, Note: This book contains fifteen photographs of Thylacines taken at this zoo.

‘Bunyips and Bigfoots In Search of Australia’s Mystery Animals’, by Malcolm Smith.

‘Out of the Shadows’, by Paul Cropper and Tony Healy.

SOME THYLACINE DVDs AND CDs

The following 1996 educational fifty-eight minute documentary DVD is available from the Mole Creek Hotel, in Mole Creek, Tasmania, and from The Tasmanian Museum and Art Gallery in Hobart, Tasmania:

'The Tasmanian Tiger', by Winning Post Productions. Note: This DVD has all know movie footage ever taken of Thylacines. It also gives some details of Thylacines in West New Guinea, where they are known as a Dobsegna be the natives of the area where they are found.

The following 1999 sixty-minute audio CD has recordings of adamant Tasmanian people describing their Thylacine sightings, since 1980, to Ned Terry, It is also available from the Mole Creek Hotel:

'Tasmanian Tiger Alive and Well.'

THYLACINE SIGHTINGS INFORMATION ETC. ON 'THE WEB'

www Tassie tiger.org, is an educational website created by Murray McAlister. It has numerous sightings listed in chronological order, a Media section containing photographs, videos, and an interesting audio radio interview that Murray had with a Victorian FM radio station. On it he talks about his Thylacine research, including his sightings of them, and his expedition to Tasmania etc.

The website www Tasmanian-tiger.com has a map of both mainland Australia and the state of Tasmania. Dots have been placed on these maps, which were produced by the Western Australian Museum, the South Australian Museum, and others, according to the map credits below each one. Each dot indicates the location of a Thylacine sighting since they officially became 'extinct' in 1936, up until 1980.

A 'Google Search' on 'Thylacine Sightings', gives a list of numerous websites detailing sightings etc.

A 'Google Images' search on 'Thylacines' gives both genuine and false images of Thylacines.

A 'Google Search' on 'Thylacine Videos' gives movies of them, and Interviews with witnesses etc.

Lightning Source UK Ltd.
Milton Keynes UK
UKOW04f1902090715

254907UK00001B/20/P

9 781925 209402